THE ART OF SYSTEMATIC TROUBLESHOOTING

"Taking the pain and time out of troubleshooting"

By Shawn A. Pinnock

"Applied Knowledge is Power"

THE ART OF SYSTEMATIC TROUBLESHOOTING

"Taking the pain and time out of troubleshooting"

By Shawn A. Pinnock

"Applied Knowledge is Power"

THE ART OF SYSTEMATIC TROUBLESHOOTING
Published through Lulu.com

Book Design and Layout by
www.integrativeink.com

TABLE OF CONTENTS

OBJECTIVES OF THIS BOOK

The objectives of *The Art of Systematic Troubleshooting* are listed below. They have been developed to help you, the reader, with often-encountered situations in troubleshooting.

Objectives:

- To explain the importance of proper mental preparation before tackling a problem
- To explain how to remain on course and avoid distractions
- To explain how to outline and define a problem
- To describe the techniques used to solve different types of problems
- To emphasize the importance of logical and systematic ways of thinking through a problem

This book deals with how to handle technical problems that are encountered in a domestic, commercial, and industrial environment, with guidelines on how to approach these problems in a logical and systematic manner, which will reduce your downtime and make your repairing, servicing, and maintenance department more effective. This book covers both the technical and mental aspects involved in dealing with problems, and it is ideal for electrical, electronic, and mechanical service technicians, maintenance, and also engineers and managers of any level who, from time to time, may deal with these problems.

INTRODUCTION

Troubleshooting is a skill that, if properly executed, can be effective in reducing your downtime. Whether it's a mechanical, electrical, or electronic system, there are certain steps that must be followed in order to be effective in troubleshooting. Troubleshooting can be thought of as the art of detecting, and one of the greatest tools you have is your mind. From experience, you will realize that correcting a problem is not difficult, but finding what is causing the problem can be a very painstaking and time-consuming exercise, and that is what we are going to deal with here: learning how to reduce the time and pain involved with finding the cause to a problem.

Effective troubleshooting requires a systematic approach and the use of methods that will yield results in a very short time. It is important to have a consistent approach to troubleshooting problems, and it is also helpful if the approach is general enough to work for a variety of problems. The troubleshooting process that will be used is called the Five-step Process:

1. Identify the problem.
2. List the possible causes.
3. Test each possible cause until the problem is found.
4. Correct the problem.
5. Verify that the problem is corrected; that is, the equipment should be working properly.

This troubleshooting process can be further broken down into two categories, the mental and the technical aspect.

In the mental aspect, the following topics will be covered:

- Humility with Confidence
- Impartiality
- Mental Blocks
- Time for a Break
- Anger
- Fear

In the technical aspect, the following topics will be covered:

- "Overworked"
- "Divide and Conquer"
- Keep Track
- How to Separate into Sections
- Substitution Test
- Intermittent Problems
- Systems

MODULE ONE
PROPER MENTAL APPROACH

"PRACTICE MAKES PERFECT"

Module Objectives:

After completing this module, you will be able to:

- *Develop the proper mental approach before tackling a problem*

- *Identify different types of mental blocks in solving a problem*

- *Learn how to deal with the different types of mental blocks*

Humility with Confidence

After years of doing troubleshooting, you will notice the importance of the words *humility* and *confidence*. They may seem to contradict each other, and you might be asking yourself, "How can one be confident and humble at the same time?" You are right; it is not easy to be both at the same time! Therefore, this is where training your mind to think differently begins.

Humility is freedom from arrogance, and in troubleshooting to find true answers, arrogance can be a big distraction. You don't want to start off on the wrong footing when trying to find answers to questions, and that's the reason for being humble. This ensures that you don't commit yourself before the facts are founded. Making claims before the facts are founded will put you in an embarrassing position. You may find yourself shifting focus away from finding the true cause of the problem and instead focus on validating your claim. It's human nature; it happens to all of us.

To prevent this from happening, you must keep an open mind, and do not make any claims to the cause of a problem until the facts are known. Instead, make a list of possible causes and examine them one by one. With **confidence** and proper examination techniques, you validate each one of the causes. Any that do stand up to examination can be disregarded, and attention should then be drawn to those that do not. After making the corrections, test the system again to make sure that everything functions properly. If it does, you have found the cause, but if it does not, you still have more testing to do.

IMPORTANT NOTE:

In troubleshooting, "YOU MAKE THE FACTS TAKE YOU WHERE THEY MAY." Do not try to influence the end result. If you do, you will never truly solve the problem!

This might better be explained by an old story I heard. This story is about two servicemen—one a bright youngster, still wet behind the ears and just out of college with his degree, and the other an older man with much experience and an equal educational background. Both men were asked the same question about the possible cause of a certain problem. The youngster, full of confidence in his knowledge and skill, gives a quick, positive answer, while the more experienced man says, "Well, I don't know! Let's check it and find out!"

He is humble enough to admit that he doesn't know all the answers right away. In addition to this, he knows from experience that the chances of a rash guess being right are awfully small. So, he stays humble and quiet until he has a chance to get the facts through performing tests. In doing so, he is able to abandon his initial guess and make another at any time. He has confidence in himself and in his ability to find the trouble and fix it eventually. So, he can afford to be humble. The youngster, on the other hand, anxious to prove himself, shows overconfidence. Often he suffers for it until he, too, learns the patience and humility with which the job should be approached.

Remember, you want to create a list of possible causes to the problem, then put these causes to the test by first doubting them and then letting the tests convince you of the cause to the problem. This was probably best explained by Ronald Reagan with the saying, ***"Trust, but verify."***

Impartiality

As you may see, developing on the concept of having an open mind is very important. In finding answers to problems, you will need to be impartial and cultivate an open mind.

Be prepared to discard any possible cause that is not validated by tests. Your objective here is to find answers. Do not make the common mistake of believing that your guess is always right—and then trying to prove that it is—but instead investigate every clue and let it take you where it will. Pay attention to everything, especially small things; they are often ignored, and most of the time, they are the cause of the problem.

To sum it up, your main concern here is to find the cause of the problem and to fix it—not to concern yourself with the politics of the cause to the problem.

Mental Block

A mental block, as the name suggests, is the inability to think in a logical and clear manner. Some of the symptoms of mental block include:

- Being stuck on one idea without any factual reason
- An inability to make sense from your tests
- The machine or system looks difficult and more complicated than it actually is
- Performing the same tests over and over
- No new ideas are flowing

The reasons for mental block usually include fatigue and working too long on a faulty machine at one time. In some cases, your mind might be preoccupied with other pressing issues, and your mind is just not on what you are doing.

The only way you will be able to think again is to get rid of this mental block, which leads to the next topic.

Time for a Break

The brain is like a muscle, and it may grow tired after thinking through a problem for a while. The solution is to take a break from the problem. There are indicators that will let you know when it's time for a break, and these indicators are:

- You are not thinking of any new approaches or ideas.
- The same conclusion keeps popping up, but it has been proven to be wrong.
- You have caught yourself just staring at the machine.
- Nothing is making sense!

Those are indications that it's time to get up, just walk away for a while, and do any of the following:

- Talk to someone—but not about the problem.
- Take a short nap.
- Watch a movie, listen to some music, read a book, or just take a walk.
- Have a drink or smoke a cigarette.
- Do anything, but just take your mind off the problem for a while.

You would be surprised by what a difference a little break can do. The mind works in two parts: the conscious and the subconscious. The conscious is the part that we are aware of when it's working, while the subconscious is the one that's working in the background, which basically functions without us being aware of it. So, resting the conscious part of the brain removes that mental block and allows for the movement of ideas between both parts.

Anger

Anger is another form of mental block, and it's something we have to keep in check. Remember, these are machines that you are dealing with, not people. Where being angry or emotional with a person might enable you to get your point across, or even resolve a problem, it's not that way with machines. Machines have no feelings, and they do not respond to anger or emotions; they respond only to logic or methods, so save your anger and emotions and remain calm—it's the only way to maintain a logical and systematic thinking process.

The next time a problem seems stubborn, don't get angry and smash or bang the machine; instead, remain calm and think of this question: *"What is it that this machine is looking for that it is not getting?"* Once you can answer that question, you will become the best of friends with the machine.

Avoid becoming angry, remain calm, and be patient. The key here is to remain focused and keep thinking in a logical and systematic manner. Remember, machines are the most patient things you can find, so think of that as a plus; it will be there waiting on you to make the right moves to allow it to work again.

Fear

Fear is probably one of the greatest mental blocks that people encounter, and the fear that I am talking about is the "Fear of failure." Ironically, this fear of failure will be your number one reason for failing, so it does you no good to begin with it.

To overcome this, you will have to change your mental approach to one of enforcing that you have the technical ability to get the job done. The approach should be one of confidence that you are capable of solving this seemingly difficult problem in a timely manner—but one must be careful of over-confidence, which may lead to simple mistakes. You must have confidence to the point that you know you will find the problem eventually through a logical and systematic process but not over-confidence!

"Confidence, if you don't control it, it will control you."

MODULE TWO
DIVIDE AND CONQUER TECHNIQUE

"UNITED WE STAND, DIVIDED WE FALL"

Module Objectives:

After completing this module, you will be able to:

- *Understand the "Divide and Conquer" technique*

- *Know when to use the "Divide and Conquer" technique*

- *Know how to apply the "Divide and Conquer" technique*

Introduction to Divide and Conquer

Systematic troubleshooting is a method of troubleshooting that is done in a logical way, which narrows the causes of a problem to a more specific area until that cause is found. By using such methods, you can reduce your troubleshooting time to a minimum and avoid troubleshooting in a circle. Instead, you can troubleshoot in a straight line, based upon understanding the information received from each measurement and having a plan of action.

The first method of troubleshooting that we will look at is what I call "Divide and Conquer."

Divide and Conquer

I like to think of this with the simple phase "Divide and Conquer" because it actually explains the principle behind this method of troubleshooting. Most systems are made up of several sections, so the idea behind this method is that you can quickly eliminate half of your time by dividing the system in half right away, with only one check. Isn't that cool! You have now reduced your area of troubleshooting in half. You then continue in this manner until you have reduced the problem to one small, specific area where you are able to execute repairs.

The tricky question here is, "When should I use this method?" The answer to that is, this method should be used when you have no idea where the cause of the problem is located within the system. In other words, you have no reason to believe it is in one section verses another, and this method works better if the system consists of sections that are configured into a series (this series configuration will be dealt with later in another section). The system is then divided into smaller sections (these sections should be meaningful sections that actually perform a specific function or duty).

Now you choose the mid point of the system, and then you make your first reading or check. Based on the result of that reading or check, right away you will know if the problem is located ahead or behind that checkpoint. In effect, you have now reduced your system in half, so you then continue with the other half, whichever it is, and cut that half into another half, and so on. Use this same procedure until you have finally narrowed the problem to a specific area, where you are now able to execute repairs.

There is the possibility that more than one cause exists, but unless you have reason to believe so, you should assume there is only one fault that exists.

To further clarify, let us look at a real life example that will help you grasp this concept fully.

Example 1

Parts Transfer Conveyor Belt

We have all been involved in a situation where a motor works properly, and then one day, we are told that this motor is constantly cutting out.

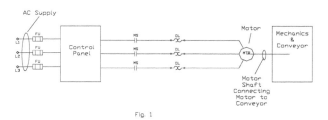

Fig. 1

Assuming we identify that there is a problem, let's go through this system. Now, the first thing we need to do is to divide this system into smaller sections:

- Electrics and Controls
- Motor
- Conveyor and Mechanics

So, right away we have three major parts that make up the system. Now, please note that we are trying to find the cause of the problem. (The motor might be burnt, but that is not

11

necessarily the cause of the problem. It might be something else that has caused the motor to be burnt in the first place, so try to avoid common mistakes such as this.)

The first thing we want to do is to divide this system in half. (We are assuming that we do not have any reason to believe the problem is located in one place verses another.) This system has electrics on one side and mechanics on the other side, with the motor connecting both of them together. To find out if this is a mechanical or electrical problem, we check to see how difficult it is to turn the motor shaft. If it is difficult to turn, it is most likely a mechanical problem. If not, it is likely an electrical problem.

If it is a mechanical problem, we can further reduce our check area to checking the mechanical section, such as:

- Bearings
- The conveyor for jams
- Chains or Belts, if any
- Mechanical overloads, like weight on the conveyor belts

If it is an electrical problem, we can further reduce our check area to checking the electrical section:

- Supply voltage
- Overload system, if it is tripping according to its specifications
- The motor, insulation, ventilation
- Slack connections

In the above example, you would be able to identify the cause of the problem in two steps. In other methods, you could have taken as much as four to six steps before identifying the cause of the problem!

These are some of the things you could check; the idea here is to give you an example of how a method like this would be used.

Let us take a look back at what was done:
(In this case, the last two items were not done, but this is the way the problem would have been resolved.)

- We identified that there was a problem.

12

- We divided the system into sections that could be involved with the problem.
- We checked each possible cause, based on the previous check.
- We corrected the cause of the problem.
- We completed a final test to verify that the system is now working properly.

Like everything else, you should look at your situation and use the method, which will best suit the situation. Sometimes these divisions are not easily done because of access to the different areas in the system. When this is so, you will have to modify your procedure to get the job done, while still keeping your troubleshooting time to a minimum.

MODULE THREE
THE OVERWORKED TECHNIQUE

"IT'S THE SQUEAKY WHEEL THAT GETS THE GREASE"

Module Objectives:

After completing this module, you will understand:

- *The meaning of the "Overworked"*

- *The importance of the "Overworked"*

- *How to apply the technique of the "Overworked"*

The Overworked

From experience, you will learn that there are some parts in a system that are more likely to fail than others. Being aware of this fact will allow you to save time in your troubleshooting process. I call this the "Overworked." As the name implies, it is the part of the system that does most of the work or the part that is working at all times. There are a few things that are common among these parts, which I have listed below. They are not in any particular order because their order may depend on the situation.

In the system, we should look at the following areas or parts that:

- Pull high current
- Absorb high power
- Carry high voltage
- May be on continuously
- Have a lot of moving, mechanical parts
- Have intermittent operations, with short off time
- May be subject to environmental conditions such as heat, dirt, and chemicals

Example 2

<u>DC Drive System</u>

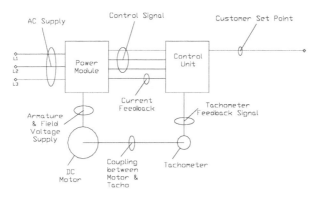

Fig. 2

15

Brief Operation of the System

The Power Module is what is called the muscle of the system. It carries the power required to turn the motor under a load. The Power Module usually consists of a silicon controlled rectifier (SCR), which converts the AC power to DC power for the motor.

The Control Unit is what is called the brain of the system; it controls the SCR in the Power Module by controlling the firing angle of the pulses that go to the gate of the SCR, which in turn controls the power to the motor. The Control Unit collects information from the motor in terms of speed, armature current, and in some cases the field current. It then compares this information with the customer set point and, depending on the difference or error of the information collected and the customer set point, it makes the correction by altering the firing angle of the SCR.

Problem

It was reported to you that the DC motor stopped turning. This is a good circumstance in which to use the "Overworked" method. This is due to the fact that this system has a big difference in power handling in the Power Module in comparison to the Control Unit. In applying this method, the following order was used in making the checks:
(Note: It is assumed that the problem has been identified before proceeding.)

- The motor and also the ventilation system
- The Power Module
- The Tachometer
- The Control Unit

The motor was first checked for the following:

- Shorts and open circuits in both the Field and Armature
- Proper ventilation

These systems were all okay, so the Power Module was checked next, where the following components were examined:

- SCR
- Diodes
- Capacitors

Two of the SCR's gates were found to be open, along with one diode, so these defective parts were replaced, and then the system was put back together, except for the motor. The motor was replaced by light bulbs for the armature and field. The reason for doing this is to make sure that everything is working properly with the Power Module and Control Unit before reconnecting the motor. The field bulb lit up perfectly, indicating that the field circuit now functions, but the armature bulb did not. The output of the Control Unit was checked, and it was found that three of the output pulse transformers were bad. These bad transformers were changed, and the system was put back together with the light bulbs. Everything worked as expected. The field bulb lit up and the armature bulb came right up to full brightness, as expected with an input from the customer set point. The bulbs were removed and replaced by the DC motor, and the final test was done. Everything went okay.

In this system, this is one of the more likely approaches to adapt in solving the problems that might arise within the system, as the level of probability of one section failing in comparison to the other is so great that you would use this method.

Now, let us look back at what was done:

- We identified that there was a problem.
- We made a list of possible causes in order of those parts most likely to fail.
- We checked each possible cause, based on the previous check.
- We corrected the cause to the problem (replaced defective parts).
- We completed a final test to verify that the system now works properly.

MODULE FOUR
"KEEPING TRACK" TECHNIQUE

"THE PEN IS MIGHTIER THAN THE SWORD"

Module Objectives:

After completing this module, you will understand:

- *The importance of keeping track*

- *What you should keep track of*

- *How to apply this technique of keeping track*

Keep Track

After gathering your initial information regarding the system that is not working properly, you should make a pencil trail of the areas that could be causing this problem in order of most likelihood. This not only outlines your possible causes but it also helps you to keep track of what is being tested, what is okay, and the results from the previous test.

In systematic troubleshooting, the idea is to reduce your area of possible causes to a smaller and smaller area through testing. In complex systems, this method helps to avoid repeat testing of the same area and the confusion this can cause.

This is a good habit to develop. Even if you are unable to keep track with pencil and paper, you should try to create a mental picture of this list of possible causes so that you will be able to eliminate your good areas. You can then move on to your next best area of possible cause. This will avoid testing in circles, which leads you nowhere.

Example 3

PLC (Programmable Logic Controller) Control System

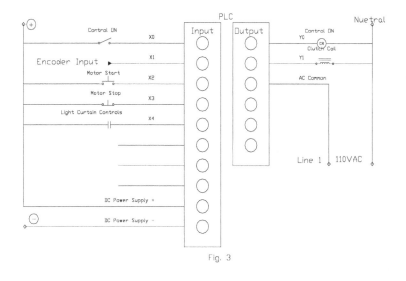

Fig. 3

Brief Operation of the System

The PLC is the base of the control system for this stamping machine, and connected in parallel with the 24VDC power supply of the PLC is an auxiliary 24VDC power supply for field sensors.

The following are connected to the inputs of the PLC:

- Control ON, which turns the control circuit on by sending 24VDC to input X0, and then the PLC output turns Y0 on, which causes the control ON relay to energize and send power to the field sensors.
- Encoder input is on X1. This unit tells the PLC the position, speed, and direction of the main shaft.
- Start main motor button is on X2. When pressed, this will start the main motor.
- Stop main motor button is on X3. When pressed, this will stop the main motor.
- The light curtain's control output goes to X4. When energized, this will cause the PLC to de-energize the clutch.

Note: For simplicity, we have left out the other sensors and switches.

Problem

The operator makes a report to the repair department outlining the following: when the control ON button is pressed, the control ON light comes on and then goes out. The main motor won't start, and the machine won't do anything.

With this information, the schematics were examined and the possible trouble areas were highlighted, as follows:

- Control ON relay
- 24VDC Power relay
- Field Sensors

Now the aim is to try to reduce this list of possible causes. The control ON relay is energized and checked for proper contact.

Proper contact appears to have been made, but after a few seconds, the control ON relay is de-energized. Upon further checks, it is discovered that the DC supply voltage disappears and comes back up after the relay is de-energized. Based on this observation, the problem was further narrowed down to one of two things happening:

- The DC power supply is overloaded when the control ON relay that powers up the field sensors is energized.
- The DC power supply is no longer capable of carrying its rated load.

This narrows the list to two possible causes, now based on the most likelihood of failure; the first thing to check for is an overload situation. In order to check the field sensors, a list of them will have to be made. Each item should be checked off once they are tested and found to be in proper working order. The list is as follows:

- Encoder
- Light Curtains and its controls
- Proximity sensors

This list is in order of those parts most likely to fail. The encoder is listed first because of its moving parts, such as bearings, couplings, and an encoder, which in general is very delicate. This is followed by the light curtain controls due to its solid state electronics, and last but not least, the proximity sensors, which have no mechanical movements and are very resilient to their surroundings.

Going down the list, the sensors were tested. First, the encoder was disconnected from the 24VDC power supply and then checked again. This time, the control ON relay stayed energized and the main motor was able to energize, which indicates that the encoder was overloading the DC power supply. The encoder was replaced and a final test of the machine was made to ensure that the new encoder was corresponding correctly to the math already programmed into the PLC.

In the above example, a track was kept of the suspected troubled areas and new notes were made as new readings were taken. This avoids the possibility of checking the same thing over and offers more information regarding the possible cause of the problem.

Now, let us look back at what was done:

- We identified that there was a problem.
- We made a list cf possible causes in order of those parts most likely to fail.
- We checked each possible cause, based on the previous check.
- We corrected the cause of the problem (replaced defective parts).
- We completed a final test to verify that the system now works correctly.

REMEMBER, KEEP TRACK AND MAKE NOTES. YOU WILL NEED THEM!

MODULE FIVE
TECHNIQUES FOR DEALING WITH INTERMITTENT PROBLEMS

"ALL THINGS ARE DIFFICULT BEFORE THEY ARE EASY"

Module Objectives:

After completing this module, you will learn:

- *The different types of intermittent problems*

- *The causes of these different types of intermittent problems*

- *How to approach these different types of intermittent problems*

- *Troubleshooting techniques for these different types of intermittent problems*

Intermittent Problems

These problems are the worst type to deal with, but they can be overcome if approached correctly. The system works well most of the time, then it suddenly stops working, and then it begins to work again the moment you try to make some checks. In most cases when the system stops working, you do not have enough time to make checks, and this becomes time consuming and annoying. But not to worry, there are ways of overcoming this type of problem more efficiently by forcing the problem to show up or by installing monitoring equipment to monitor certain test points.

Types of Intermittent Problems

Intermittent problems are generally caused by three different sources:

- Mechanical
- Thermal
- Electrical

The major characteristics of intermittent problems are the opening or shorting of circuits somewhere in the system. Intermittent shorts are usually easily found because of their obvious signs, such as overheating or traces of arcing, both visual and audio, to name a few. In this example, we will concentrate mainly on problems caused by opening, because this type of problem is much more difficult to identify.

Mechanical Intermittent

Movements of parts that open and close connections within the system by mechanical means cause mechanical intermittent problems. To explain this in simpler terms, these are some examples of intermittent problems: bad solder joints, loose or dirty connectors, worn parts, vibrations that cause conductors and joints to break, and so on.

The first thing to do is to identify whether the intermittent problem is mechanical or not, which can be done by mechanically bending, vibrating, and tapping the unit to see if the problem shows up (this is what is done to force the problem that is mechanical). After this fact is established, it is now time to try and find the section or the part that is causing this problem. This check can be done by shaking and tapping each section of the unit while checking the output for changes, and remember to take note of what section causes the greatest effect. The output can be monitored by means of an oscilloscope, a voltmeter, or anything that will give some indication of the output of the unit.

Systems that are prone to this type of problem are those that contain the following parts, or are exposed to any of the following conditions:

- Vibrations
- Connecting cables, especially moving cables
- Plugs and connectors
- Sockets
- Moving parts, such as switches, contactors, variable resistors, and so on
- Dirty environment and oxidation
- Parts that are subject to wear

Let us look at an example of a typical situation with a mechanical intermittent problem.

Example 4

Light Curtain on a Press

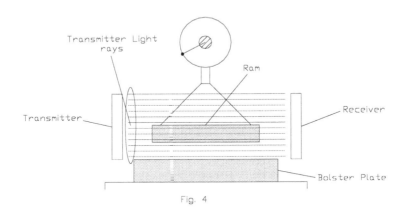

Fig. 4

Brief Operation of the System

This metal stamping Press has light curtains in front of it in order to protect the operator and equipment from injury and damage. The light curtain is made up of two parts, the transmitter and the receiver. The transmitter transmits a series of light beams, which are picked up by the receiver. If this beam is broken for any reason, it will signal the Press to stop immediately.

Problem

It was reported that every time the Press ram came down, the light curtain would stop the Press, and there was nothing breaking the light beam. The operator said that he noticed that this only happened when the ram was down, and especially with big dies. Other than that, it worked well.

From all reports, this sounds like a mechanical intermittent problem. In order to test this theory, the receiver was shaken while power was on the light curtains. Nothing unusual happened. Then the connectors to the receiver were also shaken.

Still, nothing showed up. So, for now, we can assume that the receiver is okay.

The transmitter was then shaken. The indicating lights went out, and the Press was signaled to stop. Now the theory has been proven—there is something inside the transmitter that is opening and closing, which causes the loss of power to both the receiver and the transmitter.

The transmitter was taken from the Press to do further checks. The transmitter carries the power supply for both the transmitter and the receiver, and there are circuits for the twenty-four light bulbs that transmit the light beam to the receiver. The first thing that was checked was the power supply for any loose or broken connections, and a few were found—the power transformer and the filtering capacitors. These were mounted on the printed circuit board, and because of the constant vibration of the Press and the weight of these rigid connected components, they eventually broke loose. These components were replaced, but the mounting was done in such a way as to prevent the same thing from reoccurring.

It is always important to collect as much information as possible about a problem and then evaluate that information with the actual equipment and its surroundings.

Now, let us look back at what was done:

- We identified that there was a problem.
- We created a list of the possible causes (developed theories regarding the cause of the problem).
- We checked each possible cause, based on the previous check (tested the theories).
- We corrected the cause of the problem (replaced defective parts).
- We completed a final test to verify that the system now works correctly.

Thermal Intermittent

Thermally intermittent problems are caused by the expansion and contraction of parts within the system by thermal means. This type of intermittent problem is a bit tricky, but these are normally

27

caused by overheating due to slack connections, poor ventilation, over current flow, dirty parts, and so on. You can check for thermally intermittent problems by looking for actual hot spots, overheating components, and overheating conductors.

This can be done in several ways. The easiest way is to check the temperature of the components by simply touching the suspected parts. If you have access to an infrared camera, this makes it so easy it's not funny! Simply take pictures of the different sections, and you will actually be able to see the hot spots in each section by comparing the different color gradients in the picture with the color temperature chart to determine which is above normal. After finding a part that you suspect, you can monitor the output of the unit while cooling down the suspected part, either by directing forced air to that part or by simply holding this part with your fingers! (*Be careful! If it is extremely hot, do not try cooling it with your fingers!*)

Systems that are prone to this type of problem are those that are exposed to any of the following situations or conditions:

- Loose connections
- Accumulation of dirt
- Poor ventilation
- Over current flow
- Poor solder joints

Let us look at an example of a typical situation with a thermal intermittent problem.

Example 5

AC Motor Control System

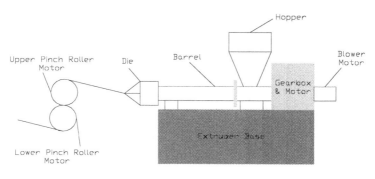

Fig. 5

Brief Operation of the System

This system carries four motors, one for each of the pinch rollers, the main extruder motor, and a blower motor for the main extruder motor. The blower motor keeps the main motor cool while running, and the main motor turns the screws in the barrel, which pushes the material from the hopper through the barrel and then through the die. This material is flattened to the desired thickness by the pinch rollers. In a fault condition, if any one of the motors should stop for any reason, they will all stop. This is for safety and material wastage reasons.

Problem

It was reported that the machine was running okay, and then it suddenly stopped. When restarted, it ran okay for a few hours and then stopped again.

Some checks were made while the machine was in operation, such as the current and voltage value of the four motors on this machine. The results were okay. Further checks revealed that the motors were protected with thermal switches, which tied into the control circuit. These switches were meant to protect the motors from overheating. From further questioning, it was learned that prior to this happening, the maintenance crew had serviced the machine. This new information suggested that the components on the machine might be okay; however, a setting might have been changed or shifted by accident.

All of the settings were checked to make sure that they were okay, and they were, so the cooling and ventilation system was checked next to make sure that there was sufficient cooling. It was found that the vent to the main motor was halfway closed, so it was opened fully, and the machine was tested again. After a few days, a follow up was done to make sure that the problem was solved and that there were no more problems. No further problems were reported on this matter.

This is a classic example of the ways in which a problem can be traced back to an event that took place before the problem started, a problem which had very little to do with the equipment itself and more to do with the people around it.

Now, let us look back at what was done:

- We identified that there was a problem.
- We created a list of possible causes.
- We checked each possible cause, based on the previous check.
- We corrected the cause to the problem.
- We completed a final test to verify that the system now works correctly.

Note: The key here was observation and questioning everything that did not seem right!

Electrical Interference Intermittent

There are two types of electrical intermittent problems, sometimes referred to as electrical interferences. These two types are:

1. Magnetic fields due to high current flow
2. Electric fields due to high voltage potential

When electrical wires of this nature are in close proximity to control wires carrying small electrical signals, they cause voltages to be induced into the control wires, which create a false signal to the receiving device. In most cases, the electric field (voltage-gradient) can be neglected, and in our case, we will be concentrating on electrical noises and induction that is due to high current flow.

Now we need to put things into their categories. When we speak of electrical noises, we are speaking of electromagnetic waves, which propagate in open air everywhere, and more so in noisy environments—for example, in manufacturing plants.

Electrical Noise

In the air, we have a lot of electrical noises, and our control wires are like antennas, which are connected to some sort of amplifier in order to amplify the low voltage signals in these wires. If our control wires are not properly protected from these electrical noises, they will pick up these unwanted signals and amplify them as well, which can cause the erratic operation of our unit. To prevent this from happening, you need to ensure that you do the following:

- Use shielded cables whenever you are dealing with small signals.
- Connect ONLY one end of the shield, preferably at the amplifier end, to a solid earth point.

If you follow these two simple rules, you should be okay where electrical noise is concerned. Now the shield works like an antenna, which picks up these unwanted signals in the air and pulls them down to earth potential, which in this case is zero volts.

Magnetic Fields

Magnetic fields are created from conductors that carry high current flow. Now, high current flow creates magnetic fields, which cause an induced voltage in nearby conductors, and the magnitude of the induced voltage is dependent on the following:

- Rate of change of the magnetic field (i.e., the rate of change of the current flow, for example, switching on and off solenoids, contactors, motors, and so on)
- The distance away from the high current carrying conductor
- The angle in relation to the high current carrying conductor
- The value of the current flow

Just as in the case of the electrical noise, if these signals are picked up, your amplifier will amplify these as well, causing the unit to behave erratically. The best bet we have against such behavior is to do the following:

- Try and maintain a maximum parallel distance between your power wires and your control wires.
- If for some reason you have to cross control and power wires, do so at 90°; this will minimize the effects.

If you follow these two simple rules, you should be okay where magnetic fields are concerned.

Now, let us look at an example of a typical situation with an electrical interference intermittent problem.

Example 6

PVC Pipe Extruder

<u>Brief Operation of the System</u>

This is a PVC pipe machine that is made up of five pieces of equipment. The first piece is called the extruder, which processes the raw material into the shape of the finished product, and the four other pieces of equipment are called the downstream equipment, which enhances the shape of the processed raw material into the final product. The first of the downstream equipment is the water trough, which is used to cool and harden the product into its final shape, while the puller is used to pull the finished product from the extruder, through the water trough, and into the saw for cutting into desired lengths. It is then moved onto the tip table for stocking.

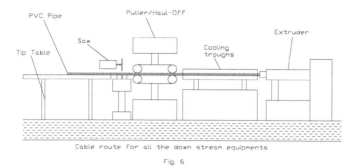

Fig. 6

The puller carries a variable speed DC drive, which is controlled from the extruded control panel, where all the other controls are also located.

Problem

The operator reports that the puller is not maintaining a constant speed; it is fluctuating.

The drive was checked for any loose connection; none was found, so the feedback circuit was looked at, which included the Tachometer and the coupling between the Tachometer and the motor shaft. The Tachometer and coupling checked out, but it was difficult to tell if the fluctuation was due to erratic signals being induced by the cables coming from the Tachometer or the Tachometer itself, so the signal from the Tachometer was replaced with a steady signal to simulate that of the Tachometer. The drive ran smoothly without any fluctuating. So then, what could have caused the Tachometer signal to be fluctuating?

The cable from the Tachometer to the drive control was checked, and it was found that the shield in the shielded cable was open from constantly rubbing against a metal edge, so the shield was no longer protecting the cables inside from electrical noise. The entire cable was replaced, and the drive ran smoothly without any further problems.

Now, let us look back at what was done:

- We identified that there was a problem.
- We created a list of possible causes (looked at the areas responsible for this function).
- We checked each possible cause, based on the previous check.
- We corrected the cause of the problem.
- We completed a final test to verify that the system now works correctly.

MODULE SIX
TEST BY SUBSTITUTION TECHNIQUE

"IT TAKES ONE TO KNOW ONE"

Module Objectives:

After completing this module, you will understand:

- *The importance of the substitution technique and safety procedures*

- *How to narrow down the system into smaller units before applying this technique*

- *How to apply the substitution technique*

Test by Substitution

The substitution of known good parts for suspected bad parts is a quick and easy way of finding a problem area, but it can also become an expensive option, if one is not careful.

If you are dealing with a system that is sub-divided into modules, which are further divided into cards, and you decide to carry spare parts in terms of modules, using the "testing by substitution" method would be quicker; however, if you wanted to repair that defective module by substitution test, you would now be required to have all the cards that make up that module. With this method, you would actually have to have a complete unit sitting there on the shelf just to satisfy this type of testing!

Next is the possibility of substituting a good part for a suspected bad part, when there is actually another defective part that is causing the suspected part to go bad. As a result, you may have now ruined your previously good part!

Knowing these facts, you should exercise caution when performing these tests in this manner, and these are some of the things that can be done to reduce costly mistakes:

- Check the outputs of the section before the suspected bad section to be substituted to make sure that its output signal or voltage is within factory specifications.
- Check the inputs to the section after the suspected bad section to be substituted to make sure that its load or next stage is not shorting or going to ground.
- Use exact replacements.

When doing this type of testing, you should still be thinking in a logical manner by trying to narrow in closer and closer to the trouble area.

Now, let us look at an example of a typical situation where we would use the method of substitution testing.

Example 7

Air Compressor Temperature Monitoring System

Brief Operation of the System

This rotary air compressor has two temperature sensors, one located at the air/oil separator and the other at the rotary screw air end. These two sensors go to an amplifier unit, which gives an alarm signal to the main control unit. When any one of these sensors goes above a certain set trip point, the main control unit will then shut down the compressor in a high temp alarm mode.

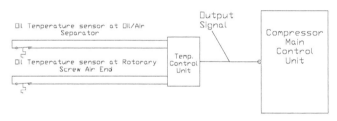

Fig. 7

Problem

After running for a few hours, the compressor shuts off on high temperature alarm mode.

The first action taken was to establish whether this alarm mode was genuine or not. The oil temperature was then checked at both the air end and also at the air/oil separator, and it was found to be below the trip limit. So, it is evident that the high temperature alarm mode is not genuine. The air/oil separator sensor was replaced with a known good sensor and tested. The problem still occurred. The sensor at the air end was then also replaced with a known good sensor, and the problem then disappeared.

From this result, we can conclude that the sensor at the air end was defective, and by using the method of substitution and logical thinking, the problem was narrowed to a single component in a relatively short time.

Now, let us look back at what was done:

- We identified that there was a problem (verified that the problem truly existed).
- We created a list of possible causes.
- We checked each possible cause, based on the previous check. (By method of substitution, the most likely thing to fail was substituted first.)
- We corrected the cause of the problem.
- We completed a final test to verify that the system now works correctly.

MODULE SEVEN
SECTIONING TECHNIQUE

"CONCENTRATE YOUR FORCES"

Module Objectives:

After completing this module, you will better understand:

- *How to identify the system in terms of each section*

- *The importance of focusing on the trouble area*

- *How to go straight to the troubled area*

Separate into Sections

While you are developing your new skills in troubleshooting, you should be able to break down a system or unit into its respective categories—its working sections and non-working sections—so as to concentrate your efforts and thinking on only the non-functional section. This technique involves the ability to segment the system or unit into smaller building blocks and then understand the function of each block. With this type of understanding, you can quickly separate what is doing its job from what is not, and then quickly determine why a specific segment may not be doing its job. This saves us time from concerning ourselves with the function of a working part, because at this stage we don't need to. What we are concerned about is the non-functioning part.

Example 8

Photoelectric Eye for a Bag-making Machine

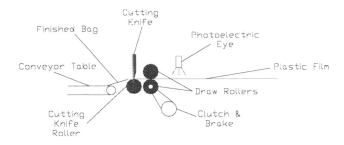

Fig. 8

<u>Brief Operation of the System</u>

Here we have a plastic bag-making machine. The machine carries two draw rollers that are driven by a clutch and brake unit. (In the newer machines, they use servo motors to do the same job.) The bag length is adjusted by cam switches. The cutting knife and the cutting knife roller, where the actual cutting takes place, are in front of the draw rollers, and this cutting knife rides on cams while the cutting knife roller is driven by the draw rollers. After cutting the bags into desired lengths, they go onto the stacking table, where they are packaged.

The photoelectric eye is located behind the draw rollers, and the sole job of this eye is to see a special mark on printed films when making printed bags. After seeing this mark, the eye sends a signal back to the controller for the clutch and brake, indicating that the mark is there. The controller then sends a signal to the clutch to de-energize and also to the brake to energize. This action stops the draw rollers, allowing for the cutting knife to cut the film at this special mark.

Problem

The operator reports that he has problems setting up the machine for printed film because the machine is not consistently cutting on the mark.

Now from practice and experience, you will be able to go straight to the area of the problem. The first thing to do is to identify what is working from what is not working. In this case, the machine ran fine when doing non-printed film; the problem was encountered when doing printed film, and the problem was inconsistency in cutting on the mark. The first check was made on the photoelectric eye to verify whether it was actually seeing the mark and sending out a signal. Next, the area on the control unit that was responsible for this function was looked at, and the optical isolator for the input signal to this control unit was found to be defective. The defective optical isolator was replaced, and the machine was tested. It was now consistent in cutting on the mark.

This might have seemed simple and easy, but this is so because of the logical and methodical process used.

Let us look back at what was done:

- We identified what was working from what was not working.
- We looked at the area responsible for the non-functioning function.
- We tested the components in the order of those most likely to fail.
- We replaced the defective part.
- We completed a final test in order to verify that the system was working properly.

MODULE EIGHT
TECHNIQUES FOR DEALING WITH DIFFERENT
SYSTEM CONFIGURATIONS

"LOOK BEFORE YOU LEAP"

Module Objectives:

After completing this module, you will be able to:

- *Show the different system configurations*

- *Outline how the input flows through the system to the output*

- *Troubleshoot the different system configurations*

Configurations of Systems

This section is mainly to provide you with an idea of how most systems are configured in terms of signal flow from input to output.

Let me start out by listing the most common types of configurations:

- Series Path
- Meeting Path
- Separating Path
- Feedback Path
- Multiplexing Path

I will now give a brief description of each, followed by an example to further clarify the concepts.

Series Path

A Series or Tandem Path is a configuration in which the input signal flows through each subsection of the system, one at a time, before getting to the output of the system. A good example of this type of configuration is the falling of dominoes that have been standing up in a straight line. The hitting of the first domino (the input) causes that domino to hit the second, and then the third, and so on in a tandem fashion until it gets to the end the system, which is the last domino (the output). The effect of hitting the first domino traveled through the system one domino at a time without bypassing any of them. This is similar to how a series system works. It has to go through each section of the system in a one by one manner until it gets to the end.

With that understanding in mind, it makes it a bit easier for a series system to be analyzed, which, by the way, is one of the easier ones to troubleshoot. The following procedure can be followed when dealing with systems such as these.

- If there is no output at all, it means one of two things:
  1. The series path is open somewhere.
  2. There is no input.
- The output is not correct.

1. The input signal got distorted along the way.
2. Something is generating a false signal within the series path.

Usually, the problem is associated with one of the two situations noted above. The idea here is that when troubleshooting a system like this, you will need to trace the input from the beginning of the system right through each subsection, until you get to the output, to find out exactly where the behavior of the input signal was not according to specification or system design.

Now, let us look at an example of how a Series Path system works.

Example 9

Motor Stop and Start Control Circuit

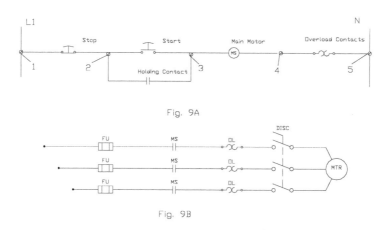

Fig. 9A

Fig. 9B

The above diagram is a stop and start circuit for an AC three-phase motor. The diagram is divided into two schematics, Fig. 9A and Fig. 9B. Fig 9A is called the control circuit, and Fig. 9B is called the power circuit. The control circuit, as the name suggests, controls the stop and start of the motor, while the power circuit is responsible for the actual connecting and disconnecting of the three-phase AC supply. The power circuit can be viewed as the muscle of the system and the control circuit as the brain.

Now the focus here is on the control circuit and how it relates to a Series Path configuration. The objective of this Series Path is for line (L1) and neutral (N) to be across contactor coil "M." In order for that to happen, the source has to flow through a number of components that are in series with the coil "M."

For the neutral to be at point "4," it has to go through the overload, normally close contact "O/L," which can be easily checked with a voltmeter. The reading between "1" and "5" should be the same as between "1" and "4," which is the same as the source voltage. That is one side of the coil taken care of. Now for the other side, which is a bit more tricky. The aim here is to get the line "L1" to point "3," but before that can happen, it has to be at point "2," which can be checked by measuring between "2" and "4." This should also be equal to the source voltage "L1" and "N." The start button is now depressed, which will bridge across "2" and "3," causing line "L1" to be present at point "3."

Providing that the coil is okay, the contactor coil "M" should now be energized, causing the contact "M" across point "2" and "3" to bridge out the start button and keep the contactor's coil energized. This contact "M" is called the holding contact.

The contactor can only be de-energized by the overload auxiliary contact or the stop button. As mentioned, this configuration satisfies that of a Series Path configuration because the current has to flow through each component.

Meeting Path

Meeting Path is a configuration in which you have two or more inputs to the system, and at some point within the system, these inputs meet. Now at the point of meeting, there are several things that can happen, depending on the design of the system. I will name a few here.

- They can add, which is called a summing point.
- They can subtract, which is called a difference or error point.
- They can route, which is called an alternative point.

These are the three most common meeting path configurations. Like everything else, in order to troubleshoot this type of system, you need to know which type of meeting path you are dealing with.

Summing Point

With the Summing Point design, the signals meeting at this point are added together, and the output is the summation of the input signals. When troubleshooting a system with this type of configuration, it is important to know your input signals because this will give you an idea of what to expect at the output of the system. Working with a system of this nature is a little more difficult because the meeting point normally involves some sort of component that takes care of the summation of the different signal values or types, such as an Op-Amps, which will not be dealt with in this book.

Unlike the Series Path, the main objective when troubleshooting this type of system is to determine if the problem is before, at, or after the meeting point, and looking at your output, meeting point, and your inputs can tell you a great deal in terms of where the problem lies.

Let us look at an example to see how a Summing Path configuration works.

Example 10

Monitoring System

This is a simple enabling or monitoring system that can be used to check the sequence of events happening. The output can always be represented by this general equation, and in this case there are three levels of output:

Input #1 + Input #2 = Output

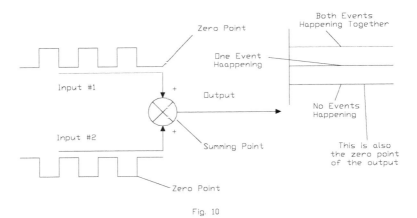

Fig. 10

1. When both events are happening at the same time, the output will look like this:

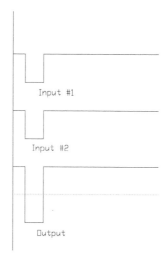

2. When one event is occurring and the other is not, the output will look like this:

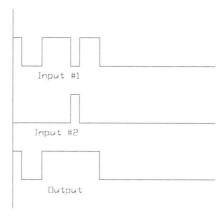

3. When no event is occurring, the output will look like this:

Input #1

Input #2

Output

Here, the Summing Point is an Op-Amp, which is connected as a summing amplifier or an adder. The output of the summing amplifier is proportional to the algebraic sum of its separate inputs.

Alternative Point

Alternative Point, sometimes called routing, is where the signals go through one common point at different times and as one output. When troubleshooting a system with this type of configuration, it is important to know your signals because it will give you an idea of what to expect at the output of the system. When working with a system of this nature, it is easier than a feedback or summing path system because the output is the same as the input, or directly proportional. As mentioned before, the main objective when troubleshooting this type of system is to determine if the problem arises before, at, or after the meeting point. This can be determined by looking at your output and your inputs at this meeting point.

Let us look at an example to see how an Alternative Point system works.

Example 11

Data Collection System

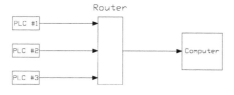

Fig. 11

This is a data collection system that is connected to three different PLCs, each located on different machines. From the PLCs through Ethernet connections, a network is set up using a router to switch the incoming data from the PLCs to the

51

computer, where the data is collected and manipulated by some sort of data collection software.

The basic function of the router is to switch the incoming data and connect it to the computer. This type of application is a good example of an Alternative Point system.

Separating Path

Separating Path is where an input is split or separated into two or more branches. These branches may be either parallel branches with the same input or filtered branches, where certain components of the input signal are taken out and sent to one branch, and a different component of the input signal is taken out and sent to another branch, and so forth, depending on how many branches. Sometimes this is called discriminative because signal selection is selective in which type of signal goes where. This type of system can be very tricky in troubleshooting because the branches can affect the input signal and also the separating path, making it difficult to determine the location of the actual problem. In such a situation, it is best to use the "Divide and Conquer" in combination with the most likelihood method discussed earlier, where you would divide the system into its units and check those most likely to fail first. Based on the result, you will know where to proceed with your checks.

Let us look at an example to see how a Separating Path system works with both parallel and filter branch types.

Example 12

Parallel Branch

DC Power Supply

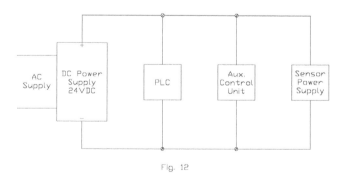

Fig. 12

In this parallel branch example, a DC power supply is used to explain this configuration. As the name suggests, the output is used to supply three separate branches:

1. Power supply for the PLC unit
2. Power supply for the auxiliary control units
3. Power supply for the field sensors

All three branches are seeing the same voltage at the same time. If a problem occurs where the supply is below its rated value, then there is a problem either with the DC supply or with the load (*in this case, one of the branches*). To find out where the fault is located, a resistive load (*based on the specified current ratings of the power supply*) can be connected to the power supply to check whether it maintains the rated voltage output. Based on the result, one of two things can happen:

1. If it does, one or more of the branches is overloading the power supply, and by process of elimination, the defective branch or branches can be found.
2. If it does not, that means the power supply is defective.

This is one of the many examples of Separating Path, where the same signal is used to feed parallel branches.

Example 13

Filter Branch

Entertainment System

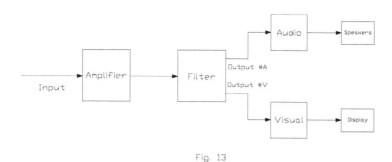

Fig. 13

In this filter branch example, an entertainment system is considered. (*This could have been a television, home theater system, etc.*) Here, the input is amplified, and after amplification, the signal is filtered. (*Here is where the filter branch takes place.*) In this case, two types of output are present. The outputs are audio and visual, and note both signals were extracted from the same input. The filter serves two purposes:

1. To filter out the noise from the input signal (*that is, the unwanted information*).
2. To separate the audio signal from the visual signal.

The audio signal then goes to the speakers and the visual signal, to the display.

NOTE: *In some systems, the audio is further filtered into high and low frequencies for the purpose of routing to different types of speakers.*

55

This is one of the many examples of Separating Path, where the same signal is used to feed filtered branches.

Feedback Path

Feedback Path is a configuration that uses a difference point, sometimes called error point, and it is where the signals meet and the difference is taken. The difference is sometimes called the error, and this error signal is actually the difference of the set value and the output value. As with any good control system, it tries to bring this value to zero. When troubleshooting a system with this type of configuration, it is important to know your input signals at this point (*which is your set point and actual point multiplied by some feedback coefficient*) because this will give you an idea of what to expect at the output of the system. Working with a system of this nature is a little more difficult because the meeting point normally involves a component that takes care of subtracting the different signal values or types.

Unlike the Series Path, the main objective when troubleshooting this type of system is to determine if the problem is before, at, or after the meeting point, and looking at your output and your inputs at this point can tell you a great deal in terms of where the problem lies.

Let us look at an example to see how a Feedback Path system works.

Example 14

Speed Controller

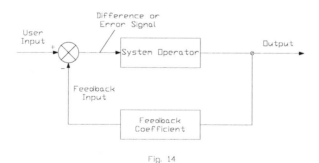

Fig. 14

57

The above figure is a block diagram of a close loop system, sometimes called a feedback system. In this example, the feedback system is a speed controller. The objective of the speed controller is to reach the speed set by the user.

The system will need to know when it has arrived where it should be. The user tells the controller where the system should be. When the user set point does not match the feedback (*the feedback in this case would be a Tachometer attached to the motor shaft*), there is a difference or error. Depending on the value of the error signal, the system will know whether it needs to accelerate, decelerate, or maintain its speed. For example, for a positive error signal, the system is behind its target speed (*called undershoot*), so it will speed up to try to reach its target speed, and for a negative error signal, the system is ahead of its target (*called overshoot*), so it will slow down to try to reach its target speed.

The system operator determines how the system responds to the error signal, such as the speed of response, the degree of overshoot, the maximum error, and the final error. The feedback coefficient is a factor by which the feedback signal will be multiplied, which can increase, decrease, or change the polarity of the signal.

This is one of the many examples of a Feedback Path, where a part of the output signal is used in comparison with the user set point to determine where the system is in relation to the user set point.

Multiplexing Path

Multiplexing Path, sometimes called switching path, is very similar to the Alternative Point that was discussed earlier. However, with multiplexing, you are able to control the path of the signal flow, while with Alternative Point, the signal flow cannot be controlled directly—in other words, the input that is present at the meeting point will be the one that goes through to the output. Knowing this makes it a bit easier to troubleshoot such a system because you can check individual inputs by switching each one on to see if there is an output (*and correct also*), and if the output is absent or incorrect, then you know something is wrong with that input.

Let us look at an example to see how a Multiplexing Path system works.

Example 15

Mode Selector

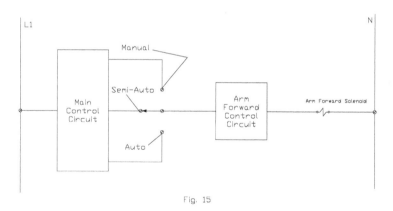

Fig. 15

The above diagram is an example of a mode selection circuit for the control of an arm's forward movement. In this mode selection circuit, there are three modes:

1. Manual
2. Semi-Automatic
3. Automatic

In the manual mode, you are able to move the arm forward manually by means of a pushbutton, while in semi-automatic mode, the machine will cycle only once when initiated. In automatic mode, the machine will run continuously until stopped or until some condition is not met within the controls.

Now, looking at this setup, one way of locating the trouble area is by doing a simple check with the mode selection button. For example, check if the arm moves in any of the mode selections. One of two things will happen:

1. If it does move, the problem lies in the main control circuit.
2. If it does not move, it could be in either place, so further checking will be needed.

Isolate the main control circuit from the arm forward control circuit, and then simulate a mode selection signal at the arm forward control circuit input. One of two things will happen:

1. If it moves, the problem is in the main control circuit.
2. If it does not move, the problem is in the arm forward control circuit.

This is one of the many examples of a Multiplexing Path, where the flow of the signal is directed to the output as desired.

MODULE NINE
THINGS TO KEEP IN MIND

"THE END JUSTIFIES THE MEANS"

Module Objectives:

After completing this module, you will be able to:

- *Emphasize the importance of the right approach in solving problems*

- *Determine how a problem should be outlined*

- *Prepare for a problem before it happens*

- *Emphasize the importance of the small things*

Final Thought

Finding the solutions to problems involves two aspects of troubleshooting:

1. The mental approach
2. The technical approach

You might be surprised to know that while most technicians, engineers, and problem solvers have the technical ability to solve problems, nine out of ten times, their reason for not finding the solution to a problem is an improper mental approach.

The proper mental approach is often overlooked, and I consider it to be the most important aspect of troubleshooting. It's like the foundation that one builds a house upon; it has to be correct for the rest of the house to stand!

The main idea behind the proper mental approach is that each problem should be examined with an open mind, and do not lean or stick to any particular results without facts to support it. When I say facts, I do not mean an opinion, a gut feeling, or even experience. All of these should only be a starting point, which should be verified by suitable tests. If the test does not support it, it should be thrown out right away, and the engineer should move on to the next possible cause.

Always maintain a logical thought process. Understand the machine as best as possible, in terms of what makes it work and what doesn't. As a problem solver, you have to be very observant; take note of everything. If it does not look, feel, or sound right, it probably isn't. So check it out and verify it; this might save you a lot of time.

Develop a systematic plan when attacking a problem. As a guide, you should do the following:

- Check that there is actually a problem, and define what it is.
- Try to understand the problem as best as possible, and make a list of possible causes. (If possible, in order of those most likely to occur.)
- Test and verify each possible cause until it is found.

- Repair or replace the possible cause, then check the machine for proper operation. If the machine operates properly, then the problem is solved, but if the machine does not operate properly, you still have a problem.
- Do a follow-up of the machine after a few hours, days, or whatever time period might be appropriate. Make sure there are no other problems and that everything is still okay.

Systematic troubleshooting actually begins before the machine breaks down. It is probably said best by John Kennedy: ***"The best time to repair the roof is when the sun is shining,"*** so it begins by becoming familiar with and understanding the manuals and the schematics for the machine. This familiarity process should also involve knowing where all the manuals and schematics are located before you actually need them.

There is another important point that I think should be emphasized, and that is the importance of simple things, or what some people call the small things. As you gain experience in troubleshooting, you will realize how true and important this concept is. In troubleshooting a problem, probably nine out of ten times the cause of the problem is something simple, and I mean really simple—so simple that you might feel embarrassed at times to tell others the true cause. Remember to be confident and double check to make sure all the dots are connected, and then leave it to the machine to prove that you were correct.

Let me give you some examples of what I have come across as being simple causes to problems that I have solved.

- Loose or disconnected common points
- Dirty sensor lenses, or parts not properly adjusted
- Loose Tachometer couplings or sensor mounting brackets
- Improper operation of equipment by the operator
- Loose connections
- Damaged insulation and vibrations
- High temperature environment
- Moisture
- Switches not turned on and plugs not plugged in

These are just a few of the simple things that are often overlooked. Remember this, the next time you are thinking of overlooking something simple: If it was major, it would be quite obvious and easily seen, but the simple stuff is not easily seen or noticed and therefore requires special skills and expertise to be noticed—and that's where you come in!

Get the small stuff out of the way because small things eventually become big.

A common statement that a lot of technicians make when trying to solve a problem is, ***"In order for me to fix this machine, I will need to make some changes because it is not possible to fix it the way it is now."*** Do not make those types of statements because what you are actually saying is, ***"I am not able to fix this machine, but let me make some changes and let's see if I can get it working."*** Think of it: this machine was working before without any problems, so why is it that some changes need to be made now in order to get it back to where it was before? To me, that does not sound like someone who knows what they are doing. ***Please don't make that mistake. If it was working before, it should be able to work again by simply finding the defective part and replacing it.*** Find the problem, solve it, then, based on your understanding of the actual problem, make recommendations on how to prevent it in the future by doing some redesigning or modifications to the system.

A final note: ***"In order for you to repair machines, you have to think like the machine."***

ABOUT THE AUTHOR

Shawn A. Pinnock has fifteen years experience in industrial control and automation systems. He started working in this field in 1989, just after graduating from high school, on a simple hydraulic press. This was his first piece of equipment, which required troubleshooting due to electrical problems. The simple combinations of different types of switches and relays that made up the controls of this unit fascinated him. Gaining the tools to analyze this simple control system from reading books and discussing it with his dad (who possesses a Master's Degree in Electrical Engineering) further increased his interest in control systems.

He then got involved in control systems for the manufacturing industry, which involved a lot of automated processes, and he began working with different types of motors, drives for motors, computer base controls, PLCs (Programmable Logic Controller), and monitoring systems. He has done extensive traveling throughout the Caribbean and South and Central America doing repairs, installations, and modifications of equipment. Becoming interested in Electrical Engineering in the area of control systems, he decided to pursue a Bachelor's Degree.

He went to Florida International University (FIU) where he gained his Bachelor's Degree in Electrical Engineering, and he has also been admitted into the membership in ***Eta Kappa Nu Association (Kappa Delta Chapter)*** for excellent scholarship and other attainments in the profession of Electrical Engineering. He is a member of the following societies:

- National Society of Professional Engineers–FL Society Chapter 50
- Institute of Electrical and Electronic Engineers

- International Society of Certified Electronics Technicians

He can be contacted by the following means:

Shawn2000_jm@yahoo.com
www.lulu.com/Shawn_Pinnock

NOTES